AF443704

El Mueble *Natural* Furniture

IDEA BOOKS

**Manufactured
by Borgia Ruano**

Natural Furniture
El Mueble Natural

© IDEA BOOKS, S.A.
Rosellón, 186, 1º-4ª
08008 Barcelona-España
Tel. 93 453 30 02 - Fax 93 454 18 95
http://www.ideabooks.es / ideabook@fonocom.es

Project Management/Dirección Producción
Juan B. Lorente Herrera

Editing/Redacción
Anna García Pascual

Artistic Design/Dirección de Diseño
Lluís Lladó Teixidó

Translation/Traducción
abc Traduccions

Photomechanic/Fotomecánica
Princeps

Printing/Impresión
Rolpress

The publisher wishes to thank to all manufacturers
who have collaborated on this project.
Los editores agradecen a todos los fabricantes
su colaboración en la realización de este libro.

2001 Edición
B-38126-2001
I.S.B.N. 84-8236-209-7
PRINTED IN SPAIN
IMPRESO EN ESPAÑA
Depósito legal: B-30287-2001

ACKNOWLEDGEMENTS / AGRADECIMIENTOS

The publishers would like to thank the following manufacturers
for their co-operation in producing this book.
Los editores agradecen a los siguientes fabricantes, su colaboración
en la realizción de este libro:

ALBOR DESIGN (ANTONIO ALMERICH)
ANFE MUEBLES
ARTEFERRO
A.T.
BORGIA RUANO
COTTAGE GOLD
DANVES
EL MARANGON
FERRANDO GUANTER
FORM 75
FRANDSEN
GIARETTA
HABUFA MEUBELEN
HALO ANTIQUES
JOHN KELLY
KEEN REPLICAS
MASSIVE OAK (DE PAAL FORNITURE)
MYOC
NAFARROA
REPRODUCCIONES & DECORACIÓN
R. JUSTE REQUENA
TALGÖ
TORONDELL
VIERHAUS
ZEITRAM

I N T R O D U C T I O N

This book is dedicated to some items of furniture which, as well as for their captivating beauty and refined style, have been chosen for the quality of materials, comfort and beauty of their simple but harmonious and painstakingly produced forms.

This does not mean they lack a specific style although, generally speaking, they are not the most elaborated items. They are inspired by the first emigrants (pilgrims), in the most austere English style such as ship furniture and in the provincial French styles, Nordic furniture or from the mountain regions, making the book more up to date and really essential for choosing furniture for country mansions or for the homes of people who prefer authentic quality, exempt from pretentious fantasies.

These manufacturers and artisans select the material for their work from the best wood, taking great care in terms of its quality and origins: woods that are replenished at the same rate as they are cut down, thus avoiding damaging the balance of a delicate environment, and so necessary for the health of our much threatened planet.

The wood obtained from the forest is selected and suitably stored for the correct time (depending on the type of wood) so that it dries naturally.

It is then regulated and unified in chambers where it acquires the perfect level of humidity for panels and joints that are stable and firm.

Naturally, the climate and environmental conditions of the home in which the furniture is to be installed is of fundamental importance for its beauty and preservation.

Today, however, the fact that nearly all homes are suitably air-conditioned in winter and summer (and with humidity control at the correct levels), these items of furniture are destined to have a long and satisfying life that will last for generations.

By working mainly with solid wood, these industrialists-cum-artisans use state-of-the-art technology in order to take full advantage of the material and the best traditional procedures for the assembly and finishing of these items of furniture, in which a centuries-old cutting technique provides a grain (almost always on the straight) forming the finest of adornments: the verification of a natural and "living" material.

They are items of furniture that require the work of skilled and experienced carpenters, as well as cabinetmakers who know the techniques handed down over centuries of tradition.

They mainly use joints and pins but also quality ironwork and lock work in just the right proportions.

This is a book specialising in this type of natural furniture (for the co-operation, design and materials) which will be of great use to decorators, designers, interior designers and furniture sellers or to the amateur who wishes to knowingly choose the best furniture for their home.

INTRODUCCIÓN

Este libro está dedicado a unos muebles que además de cautivar por su belleza y depurado estilo, han sido elegidos por la calidad de los materiales, la comodidad y belleza de unas formas sencillas no exentas de armonía y una esmerada elaboración.

Esto no quiere decir que carezcan de un estilo determinado aunque, generalmente no son de los más elaborados. Están inspirados en los primeros emigrantes ("peregrinos"), en los más sobrios ingleses como los muebles de barco y en los estilos provinciales franceses, muebles nórdicos o de las regiones montañosas, esto hace que el libro sea de la mayor actualidad y realmente imprescindible para elegir el mobiliario de mansiones campestres o las viviendas de personas que prefieren la calidad auténtica, exenta de pretensiosas fantasías.

Estos fabricantes y artesanos seleccionan los materiales de sus obras entre las mejores maderas teniendo cuidado de su calidad y procedencia: bosques que renuevan en la medida en que se explotan, evitando así perjudicar el equilibrio de un medio delicado y muy necesario para la salubridad de nuestro planeta tan amenazado.

Las maderas obtenidas en el bosque, se seleccionan y almacenan adecuadamente y durante el tiempo conveniente (difiere según el tipo de madera) para su secado natural que luego se regula y unifica en cámaras donde adquieren el grado de humedad perfecto para que tableros, juntas y ensambles sean estables y firmes. Naturalmente, el clima y las condiciones ambientales de la vivienda en que se instalen los buebles, tienen una importancia fundamental para su belleza y conservación, pero, en la actualidad, en que casi todas las viviendas están adecuadamente climatizadas en invierno y verano (y con regularización de la humedad a niveles correctos)

estos muebles están destinados a un largo y satisfactorio uso durante generaciones.

Al trabajar principalmente en madera maciza, estos industriales desdoblados en artesanos, utilizan alta tecnología para un mejor aprovechamiento del material y los mejores procedimientos artesanales para el montaje y acabado de estos muebles en los que el corte según una tradición de siglos proporciona el dibujo de un veteado (al hilo casi siempre) que constituye en sí el mejor adorno: la constatación de un material natural y "vivo".

Son muebles que requieren el trabajo de hábiles y experimentados carpinteros así como ebanistas conocedores de unas técnicas obtenidas a través de siglos de tradición.

Se utilizan principalmente ensambles y clavijas, pero, también, herrajes y cerraduras de calidad sin más protagonismo del necesario.

Se trata de un libro especializado en este tipo de mobiliario natural (por la elaboración, el diseño y los materiales) que resultará de gran utilidad a decoradores, diseñadores, arquitectos interioristas y comerciantes de muebles o al aficionado que desea elegir con conocimiento de causa el mejor mobiliario para su vivienda.

BORGIA RUANO works with metal and wood, combining them with great originality. On the right: chair and armchair ref. 808. Carved Victorian console table model TV1D-M and coffee tables models CTV-M and CTO-M from KEEN REPLICAS.
BORGIA RUANO trabaja metal y madera y los combina con gran originalidad. A la derecha: silla y sillón ref.808. Consola victoriana tallada mod.TV1D-M y mesitas de centro mod. CTV-M y CTO-M de KEEN REPLICAS.

On the left: furniture with cane decoration ref. 819, bedside table and classical small armchair ref. 789 and 724. Below, on the two pages, five variations of drawers and wardrobes ref. 805c, 805i, 805b, 805c and 805a from BORGIA RUANO. Centre: coffee tables in solid mahogany model REGENCIA OCT60-M and OCT120-M, both from KEEN REPLICAS.
A la izquierda: mueble con decoración de cañas ref.819, velador y clásico silloncito ref.789 y 724, abajo, en las dos páginas: cinco variaciones sobre cajones y armarios ref.805e, 805t, 805b, 805c y 805ª de BORGIA RUANO. En medio: mesitas de centro en caoba maciza mod.REGENCIA OCT60-M y OCT120-M, ambas de KEEN REPLICAS.

El Mueble
Natural **Furniture**

On this double page,
ANFE MUEBLES
presents a bedroom in
which a deliberate
rustic style is joined
with neo-classical
carvings, turnings and
forging. The soft
natural shine of the
finish stands out here.
En esta doble página,
ANFE MUEBLES
presenta un dormitorio
en el que se aúna un
estilo voluntariamente
rústico, con tallas
neoclásicas, torneados
y forja. Destaca el
suave brillo natural del
acabado.

*This series number 18 is
produced in skidded walnut
and is made up of a bedside
table ref. 965, bed (150 cm),
chest of drawers with five
drawers ref. 970, framed
mirror (jewellery box in the
base) ref. 976 and a
wardrobe with four doors and
mirrors on the two central
ones ref. 984.*
*Este conjunto n°18 está
realizado en nogal patinado y
se compone de mesita de
noche ref.965, cama (150cm)
ref.957, cómoda con cinco
cajones ref.970, espejo
enmarcado (joyero en la
base) ref.976 y armario de
cuatro puertas con lunas en
las dos centrales ref.984.*

A writing desk is a highly decorative item of furniture as well as being useful and necessary. It can be placed in a bedroom, living room or study. There are a series of classical styles and FORM 75 supplies several attractive models.

Un escritorio es un mueble altamente decorativo pero, además, útil y necesario. Se puede instalar en un dormitorio, en la sala de estar o de trabajo. Hay una serie de formas clásicas y FORM 75, propone varios atractivos modelos.

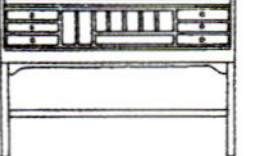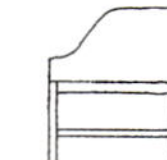

11 W/B 146 D/T 80 H 105 cm

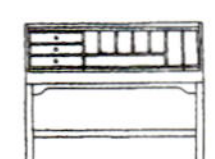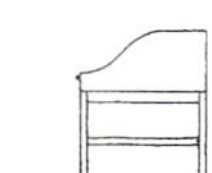

12 W/B 111 D/T 80 H 102 cm

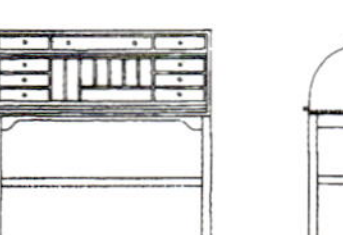

20 W/B 112 D/T 60 H 112 cm

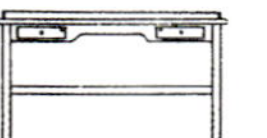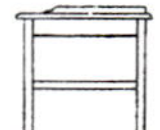

90 W/B 130 D/T 70 H 72 cm

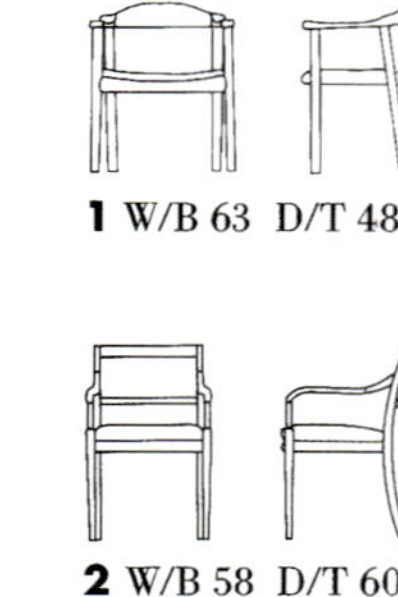

1 W/B 63 D/T 48 H 75 cm

2 W/B 58 D/T 60 H 83 cm

On the left, open writing desk
ref. W/B111 and chair ref.
W/B63, another model ref.
W/B146 and chair ref.
W/B58. On the previous
page, above: with folding top
ref. W/B90 and chair ref.
W/B63. Below: with blind
(open and closed) ref.
W/B112. Details and plans
with measurements.

*A la izquierda: escritorio
abierto ref.W/B111 y sillón
ref.W/B63, otro modelo
ref.W/B146 y asiento
ref.W/B58. En la página
anterior, arriba: con tapa
abatible ref.W/B90 y sillón
ref.W/B63. Debajo:
con persiana (abierta y
cerrada) ref.W/B112. Detalles
y esquemas con las medidas.*

HALO ANTIQUES presents furniture in solid wood: dinner table with dovetailed top and shaped legs, model WPLAT, coffee table with drawers and fluted top, model WOGTC. Beech in chair model BRUGES and armchair model PROVENCE CARVER.

HALO ANTIQUES presenta muebles en madera maciza: mesa de comedor con tablero machihembrado y patas torneadas mod.WPLAT, mesa de centro con cajones y tableros acanalados. mod.WOGTC. Madera de haya en la silla mod.BRUGES y sillón mod.PROVENCE CARVER.

On the following page, above: sideboard with upper glass cabinet, model WDA. Below: sideboard with three doors and uncovered shelves on the upper part, model WPI.AD, auxiliary table with draw and shaped legs, model WPI.AT and bookshelf, model II.
En la página siguiente, arriba: aparador con vitrina superior mod.WDA. Abajo: aparador con tres puertas y estantes descubiertos en la parte superior mod.WPLAD, mesita auxiliar con cajón y patas torneadas mod.WPLAL y librería mod.H.

El Mueble
Natural **Furniture**

JOHN KELLY designs and produces the J1 collection in cherry and hazelnut wood, finished with three hand-painted layers of oil and wax. Above: bedside table (three drawers), office desk (drawers in the side sections) and dining room.
JOHN KELLY diseña y produce la col.J1 en madera de cerezo y avellano, acabado con tres capas a mano de aceite y cera. Arriba: mesita de noche (tres cajones), mesa de despacho (cajones en los cuerpos laterales) y comedor.

On the next page: a living room atmosphere with furniture from the J1 collection. All these items of furniture combine an almost craftsman-like style with modern design and the functions are suited to modern needs: ergonomics, capacity, etc.
En la página siguiente: un ambiente de sala de estar con muebles de la misma col.J1. Todos estos muebles unen una elaboración casi artesanal con un moderno diseño y las prestaciones adecuadas a las espectativas actuales: ergonomía, capacidad, etc.

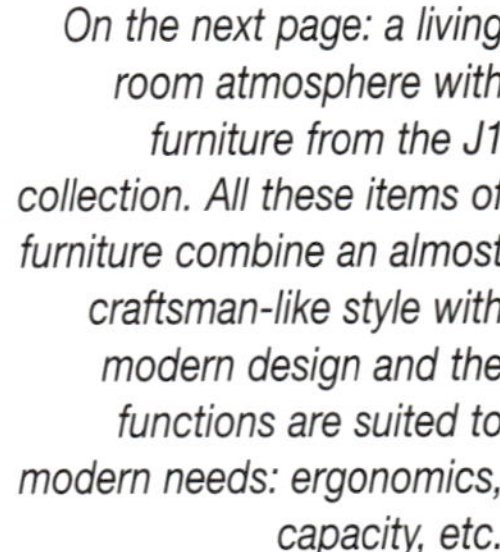

A.T. presents the N.a.K collection, made with solid pine with a wax varnished finish OR138 (on order: water lacquered or aged). On this page: glass cabinet with drawer on the lower part, model GARDA 116 x 45x 201 cm.

A.T. presenta su col.N.a.K. realizada en madera maciza de pino con acabado barnizado a la cera OR138 (a petición: lacado al agua o envejecido). En esta página: vitrina con cajón en la parte inferior mod.GARDA 116x45x201cm.

The drawers slide out perfectly thanks to the zinc lining. On the following page: a sideboard of the same characteristics, with three drawers and three doors, model GARDA, two wide flutes on the front angles, 186 x 52 x 101 cm.

Los cajones se deslizan perfectamente gracias al recubrimiento de zinc. En la página siguiente: un aparador de las mismas características, con tres cajones y tres puertas mod.GARDA, dos gruesas estrías en los ángulos frontales y 186x52x101cm.

Furniture from BORGIA RUANO: simple lines but with original details and a very natural concept of beauty and functionalism. On the right: auxiliary tables ref. 835. Below: office desk ref. 1164 and ARTEFERRO shelving ref. 1173/C.

Muebles de BORGIA RUANO: líneas sencillas, pero con originales detalles y un concepto muy natural de lo bello y lo útil. A la derecha: mesitas auxiliares ref.835. Abajo: mesa de despacho ref.1164 y estantería ref.1173/C de ARTEFERRO

On the adjoining page: front view of a mirror and chest of drawers (with doors) ref. 787 and 831 from BORGIA RUANO. Below: idea by ARTEFERRO, office desk ref. 2145/B and writing desk with upper body and folding top ref. 1184.

En la página contigua: vista frontal de un espejo y cómoda (con puertas) ref.787 y 831 de BORGIA RUANO. Abajo propuesta por ARTEFERRO, mesa de despacho ref.2145/E y escritorio con cuerpo superior y tapa abatible ref.1184.

DANVES items of furniture are faithful reproductions of Spanish colonial style furniture, with aged wood and the special "wax moth" finish. This double page shows a dining room based on the HACIENDA collection.

Los muebles de DANVES son fieles reproducciones de muebles de estilo colonial español, con maderas envejecidas y el peculiar acabado "polilla a la cera". En esta doble página presenta un comedor a base de la col.HACIENDA.

The series is made up of a sideboard with eight drawers ref. 511, rectangular table ref. 510, sideboard with display cabinet of shelves ref. 502, in the background, an auxiliary table ref. 517, chairs and armchairs, model CORAZÓN, both with cat's tail seating ref. 529 and 530.

El conjunto se compone de aparador con ocho cajones ref.511, mesa rectangular ref.510, aparador con expositor de estantes ref.502, al fondo, mesita auxiliar ref.517, sillas y sillones mod.CORAZÓN ambos con asiento de anea ref.529 y 530.

FERRANDO GUANTER has created his COLONIAL AMERICANO collection, following the guidelines of this style: an English influence with slightly more Baroque details. The quality of the material and the perfect execution ensure the final result.

FERRANDO GUANTER ha creado su col.COLONIAL AMERICANO siguiendo las pautas de ese estilo: influencia inglesa con detalles algo más barrocos. La calidad del material y perfecta ejecución aseguran el resultado.

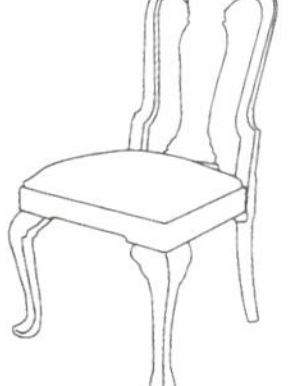

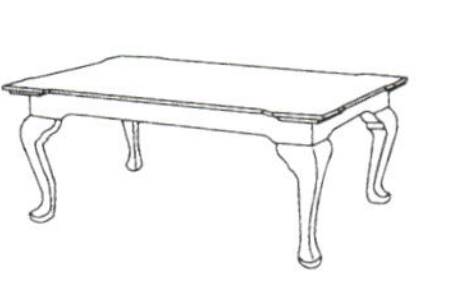

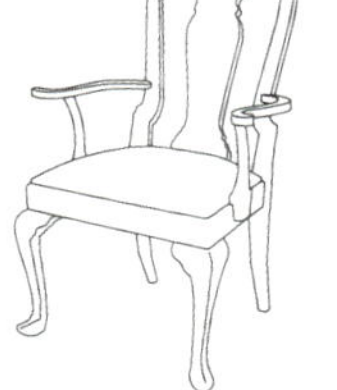

In the plans one can appreciate details and measurements of the furniture shown in these settings: the first is completed with a worktable and the other with part of the dining room (elements from the same collection). Detail of the high chest of drawers, model 46.

En los esquemas se aprecian detalles y medidas de los muebles presentados en estos ambientes de salón: el primero se completa con una mesa de trabajo y el otro con la parte del comedor (elementos de la misma colección). Detalle de la cómoda alta mod.46.

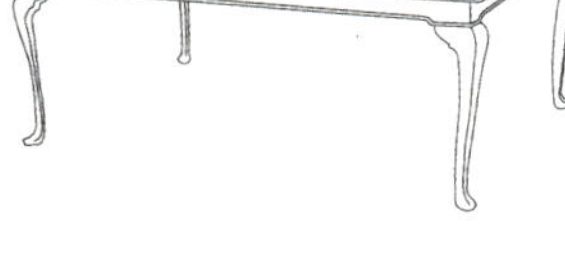

COTTAGE GOLD produces solid wood furniture (Oregon pine) which will be become tomorrow's antique collector's pieces, which does not mean that they are not designed for current use. Right: bedroom with canopy bed.
COTTAGE GOLD produce muebles en madera maciza (pino Oregon) que resultarán ser los muebles de anticuario del día de mañana, lo que no impide que estén pensados para utilizarse actualmente. Derecha: dormitorio con dosel.

Left: chest of drawers with triple mirror. Above: dining room with three types of auxiliary furniture.
Next page, above: armchairs and coffee tables (two sizes), dining room with glass cabinet sideboard and corner cupboard, full bedroom suite and round table with chairs, model VOLKSWYN.
Izquierda: cómoda con triple espejo, arriba: comedor con tres tipos de mueble auxiliar.
Página siguiente, arriba: sillones y mesita de centro (dos medidas), comedor con aparador vitrina y rinconera, dormitorio completo y mesa redonda con sillas mod. VOLKSWYN.

REPRODUCCIONES & DECORACIÓN, as the name indicates, work in solid mahogany, producing exact copies of furniture, mainly of different English styles. Step chair (open and closed) ref. SE006.
REPRODUCCIONES & DECORACIÓN, como su nombre indica, realiza en caoba maciza, copias exactas de muebles, principalmente de diversos estilos ingleses. Silla escalera (abierta y cerrada) ref.SE006.

Above and below: Victorian sideboards ref. SI002 and SI004. On the next page: Victorian bookcases of two and four sections ref. LB001 and LB003, folding table (open and closed) ref. MP050 and two bookcases: with drawers ref. LB012A and only with shelves ref. LB016.
Arriba y abajo: aparadores victorianos ref.SI002 y SI004. En la página siguiente: librerías victorianas de dos y cuatro cuerpos ref.LB001 y LB003, mesa plegable (abierta y cerrada) ref.MP050 y dos librerías: con cajones ref.LB012A y, sólo con baldas ref.LB016.

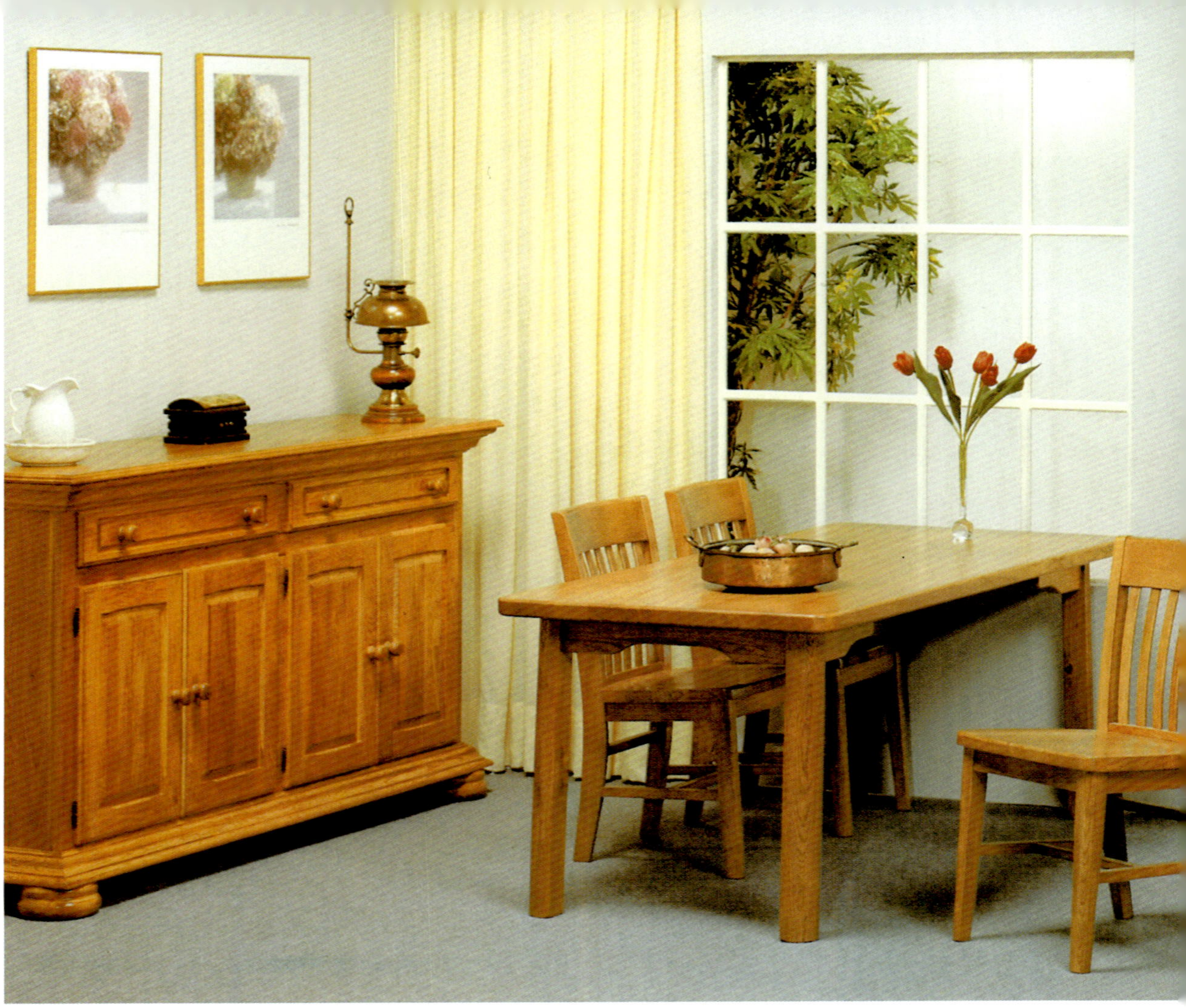

The beauty of the solid oak wood of these items of furniture from MASSIVE OAK are complemented by a traditional design and finish. Above: sideboard and display cabinet, model MICHELLE, table model GERARD and chairs model AD.
La belleza de la madera maciza de roble de estos muebles de MASSIVE OAK, se complementa con un diseño y acabado tradicional. Arriba: aparador y vitrina mod.MICHELLE, mesa mod.GERARD y sillas mod.AD.

Alongside these lines: wardrobe, dressing table with three-panelled mirror, bed and bedside table, model HANS. On the right, in column: chest of drawers for TV and video model MARTIN, sideboard, model MICHELLE and other auxiliary items of furniture.
Junto a estas líneas: armario ropero, tocador con espejo de tres hojas, cama y mesita de noche mod.HANS. A la derecha, en columna: consola para TV y vídeo mod.MARTIN, aparador mod.MICHELLE y otros muebles auxiliares.

On the left: original glass display cabinet and chest of drawers with doors and inner draw from BORGIA RUANO. KEEN REPLICAS produces furniture in solid mahogany which are exact copies of the best designs in classical furniture. Below: colonial style sideboard ref. CFAC2-M and bedside tables ref. BFBT-M and FBT-M.

A la izquierda: original cuadro vitrina y cómoda con puertas y cajonero en el interior, de BORGIA RUANO. KEEN REPLICAS produce muebles en caoba maciza que son copias exactas de los mejores diseños del mueble clásico. Abajo aparador de estilo colonial ref.CFAC2-M y mesitas de noche ref.BFBT-M y FBT-M.

The ironwork and handles of
KEEN REPLICAS are in
traditionally crafted solid cast
brass. On the left: low
Victorian sideboard ref.
SBV-M and, below, canopy
bed with carved and shaped
posts in Queen Anne style
ref. QBC-M.
Los herrajes y tiradores de
KEEN REPLICAS son de
latón macizo fundido
artesanalmente.
A la izquierda: aparador
victoriano bajo ref.SBV-M y,
abajo, lecho con dosel de
columnas talladas y
torneadas estilo Reina Ana
ref.QBC-M.

Furniture manufacturing at EL MARANGON is either a manual or advanced technology process that aims to achieve perfection as the final result. On the right: bookcase with glass display ref.1164 in solid cherry or walnut.

La fabricación de los muebles de EL MARANGON es un proceso de alta tecnología o manual que pretende, únicamente, la perfección del resultado final. A la derecha: biblioteca con vitrinas ref.1164, en madera maciza de cerezo o nogal.

Above: table ref.1410,
alongside these lines: glass
display bookcase ref. PC6
and chair ref. 996CV. On the
previous page: chest of
drawers ref. 752C, armchair
ref. 967CT, bed ref. 659C16,
small table ref. 763C and
wardrobe ref. AC6LA,
bookcase ref. SC1, chest of
drawers ref. 1706 and small
tables ref. 758/789/760C.
Arriba: mesa ref.1410, junto a
estas líneas: biblioteca vitrina
ref.PC6 y silla ref.996CV.
En la página anterior:
cómoda ref.752C, sillón
ref.967CT, cama ref.659C16,
mesita ref.763C y armario
ref.AC6LA, biblioteca
ref.SC1, cómoda ref.1706 y
mesitas ref.758/789/760C.

R. JUSTE REQUENA presents this dining room that combines the rustic (structure of the table) with the most modern (glass top), the classical (leg carvings) and the comfort of their upholstery.

R. JUSTE REQUENA presenta este comedor en que se aúna lo rústico (estructura de la mesa) con lo más moderno (tablero de vidrio), lo clásico (tallas de las patas) y la comodidad del tapizado de las mismas.

Chair ref. 7560 46 x 51 x 117 cm, table ref. 28 164 x 92 x 75 (without glass), armchair ref. 7561 63 x 53 x 117 cm, sideboard ref. 329/3B 203 x 47 x 97 cm, with upper body ref. 329/3 203 x 47 x 210 (total height). Lamp and wall light in cast iron from the same manufacturer.

Silla ref.7560 46x51x117cm, mesa ref.28 164x92x75 (sin cristal), sillón ref.7561 63x53x117cm, aparador ref.329/3B 203x47x97cm, con cuerpo superior ref.329/3 203x47x210 (altura total). Lámpara y aplique en hierro forjado del mismo fbricante.

El Mueble *Natural* **Furniture**

In this bedroom, as in other productions in the COMODÍN collection from ALBOR DESIGN, a personalised solution has been sought based on modular elements in solid wood, carefully joined and finished.

En este dormitorio, como en otras realizaciones de la col.COMODÍN de ALBOR DESIGN, se ha buscado una solución personalizada a base de elementos modulares en madera maciza, cuidadosamente unidos y acabados.

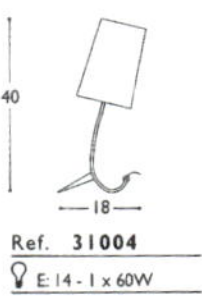

Ref. **31004**

E: 14 - 1 x 60W

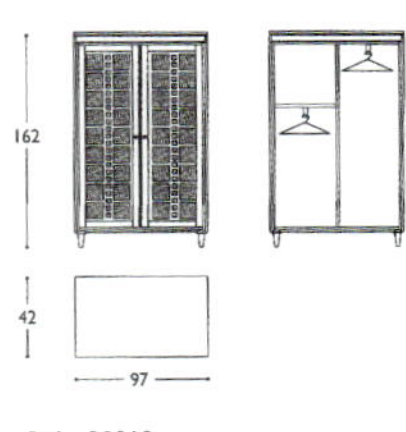

Ref. **38012**

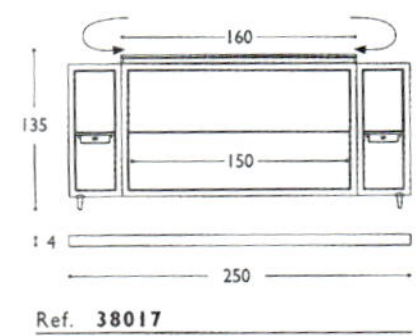

Ref. **38017**

The strict simplicity of lines and the compact design, such as the head that incorporates some small drawers as bedside tables, make up a fantastic idea for the better use of space. References and measurements of each element are shown on the plans.

La estricta simplicidad de líneas y el diseño compacto, como en el cabezal que incorpora unos cajoncitos como mesitas de noche, constituyen una gran idea para una mejor utilización del espacio. En los esquemas se señalan referencias y medidas de cada elemento.

FRANDSEN presents a dining room from its BU-NIIK collection in solid beech and with a very modern and original design. The measurements can be seen in the plans and below, another optional glass cabinet with two doors and windows on the sides ref. 76.

FRANDSEN presenta un comedor de su col.BU-NIIK en madera maciza de haya y original diseño muy moderno. En los esquemas se aprecian las medidas y, debajo, otra vitrina opcional con dos puertas y cristales en los laterales ref.76.

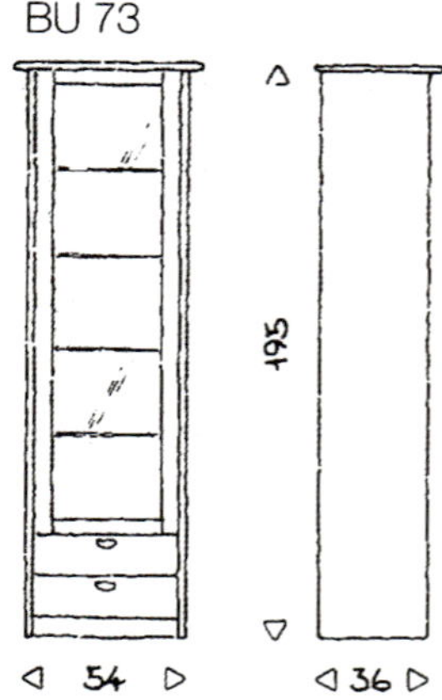

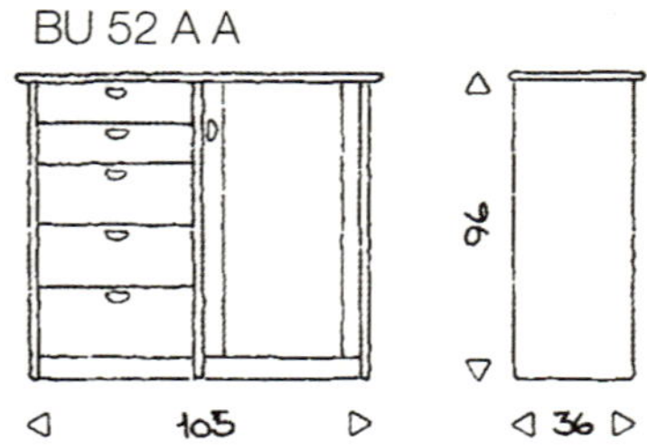

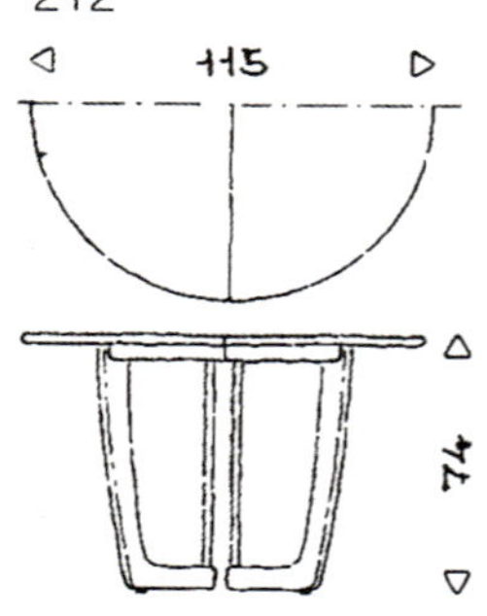

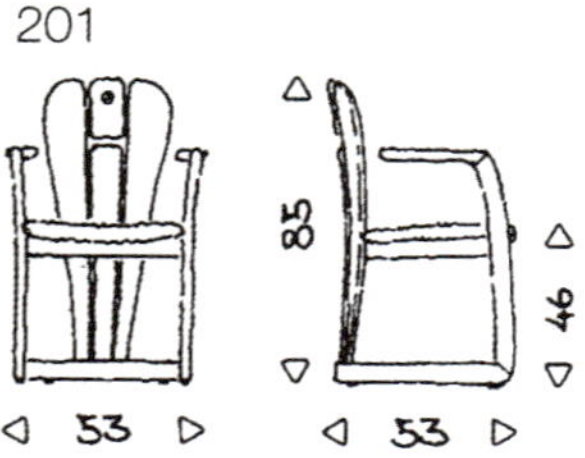

The photograph shows the wide-based armchairs and an original back ref.201, glass cabinet ref. BU73, sideboard ref. BU52AA ands round table with legs that meet in the centre in a cross as a base ref.212. An ambivalent solution that is suitable for both town and country.

En la fotografía destacan los sillones de amplia base y original respaldo ref.201, vitrina ref.BU73, aparador ref.BU52AA y mesa redonda con patas que se reúnen en el centro con una cruz como base ref.212. Solución ambivalente apta para campo y ciudad.

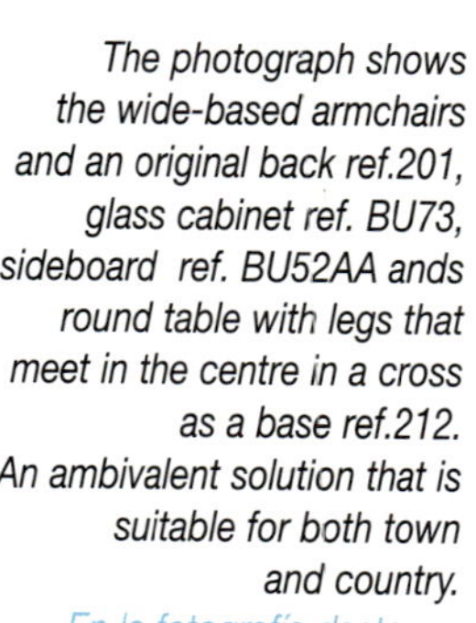

The solid wood furniture of MYOC is made up of enriching elements for decorating dining rooms. Lounges, bedrooms, etc. On this page: landscaped chest of drawers, two for the inset basin and a hammock.

Los muebles de madera maciza de MYOC constituyen elementos enriquecedores para la decoración de comedores, salones, dormitorios, etc. En esta página: cómoda apaisada, dos para lavabo encastado y hamaca.

Above these lines: three combinations of glass cabinet and sideboard ref. 10, ref. 12 and ref. 18, stool ref. 12 and high chair for bar ref. 10. On the previous page: chests of drawers ref. 22, ref. 21, ref. 21 (with basin and mirror) and a sun lounger ref. 100.

Sobre estas líneas: tres combinaciones de vitrina y aparador ref.10, ref.12 y ref.18, taburete ref.12 y silla alta para barra de bar ref.10. En la página anterior: cómodas ref.22, ref.21, ref.21 (con lavabo y espejo) y una tumbona hamaca ref.100.

The furniture from BORGIA RUANO surprises because, on being simple, they always have something new and exclusive. Below: bedroom where the head stands out with the metal embedded into the wood, chest of drawers and bedside table in straight lines. Next page: dining room combining steel, raffia and wood and a shelf.

Los muebles de BORGIA RUANO sorprenden porque, siendo sencillos, siempre tienen algo nuevo y exclusivo. Abajo: dormitorio en el que destaca el cabezal con el metal incrustado en la madera, cómoda y mesita de líneas rectas.

Página siguiente: comedor que combina acero, rafia y madera y una estantería.

Produced by ARTEFERRO and over these lines: bench in solid wood and decorated ref. 2745, small table with drawer and flap decorated with fretwork marquetry ref. 2429 and framed mirror ref. 1504. On the following page, below: umbrella stand ref. 2000, bench with drawer ref. 1121, small table, console table and mirror ref. 2379, 2007 and 1179.

Producidos por ARTEFERRO y sobre estas líneas: banco en madera maciza y decorada ref.2745, mesita con cajón y faldón decorado con marquetería calada ref.2429 y espejo enmarcado ref.1504. En la página siguiente, abajo: paragüero ref.2000, banco con cajón ref.1121, mesita, consola y espejo ref.2379, 2007 y 1179.

El Mueble
Natural **Furniture**

Dining room set with
angled bench,
rectangular table with
the traditional drawer,
chairs and sideboard,
all produced by A.T. as
part of their N.a.K.
collection in solid pine
and with a waxed finish
(natural shade).
Conjunto de comedor
con banco de ángulo,
mesa rectangular con el
tradicional cajón, sillas
y aparador todo ello
elaborado por A.T.
dentro de su col.N.a.K.
en madera maciza de
pino y acabado
encerado (tono natural).

*Table, model GARDA with
two optional sizes (145 x 185
cm or 185 x 185 cm),
sideboard with upper glass
cabinet (also from the
GARDA model), measuring
130 x 52 x 200 cm.
The shaped legs stand out,
as do the flutes and subtly
carved forms.*
*Mesa del mod.GARDA con
dos medidas opcionales
(145x185cm ó 185x185cm),
aparador con vitrina superior
(también del mod.GARDA)
mide 130x52x200cm.
Destacan los torneados de
las patas así como las estrías
y formas discretamente
recortadas.*

The choice of different models of chairs or tables creates completely different atmospheres. HABUFA presents a table with corner bench from the GLASGOW collection ref. 894 and two small coffee tables from the FURU collection: round ref. 2106 and square ref. 2130.

La elección de diferentes modelos de sillas o mesas crea ambientes completamente distintos. HABUFA presenta mesa con banco de esquina col.GLASGOW ref.894 y dos mesitas de centro col.FURU: redonda ref.2106 y cuadrada ref.2130.

On the lower part of the page, table, model ROBERT and shaped chairs. On the following page: chairs from the FURU collection: ref. 4375, ref. 4370, ref. 4365 and ref. 4334, small table with drawers ref. 2107 and more chairs ref. 4343, 4377 and 4371, all in whitened solid pine.
En la parte baja de la página, mesa mod.ROBERT y sillas torneadas. En la página siguiente: sillas de la col.FURU: ref.4375, ref.4370, ref.4365 y ref.4334, mesita con cajones ref.2107 y más sillas ref.4343, 4377 y 4371, todas en pino macizo blanqueado.

In the MASSIVE OAK furniture, the main feature is the beauty of the solid European oak, suitably selected and treated before reaching the workshop where it is produced and assembled by hand. Above: dining room, model MARTIN.

En los muebles de MASSIVE OAK, lo principal es la belleza de la madera maciza de roble europeo, adecuadamente seleccionada y tratada, hasta llegar al taller donde se producen y montan a mano. Arriba: comedor mod.MARTIN.

Alongside these lines: bed, model NIKKI, chest of drawers, dressing table and bedside table, model CARINA, mirror, model EMMIE. On the following page, above: table and furniture made up of several units (combination E). Below: dressing tables (or chests of drawers), models CARINA and HANS.

Junto a estas líneas: cama mod.NIKKI, cómoda, tocador y mesita de noche mod.CARINA, espejo mod.EMMIE. En la página siguiente, arriba: mesa y mueble formado por varias unidades (combinación E). Abajo: tocadores (o cómodas) mod.CARINA y HANS

The aim of VIERHAUS is to produce traditional rustic furniture, but with high quality solid woods (maple, birch, beech, oak, poplar, cherry and different varieties of pine), taking great care in the work and finish.

El objetivo de VIERHAUS es conseguir un mueble tradicional campestre, pero con maderas macizas de alta calidad (arce, abedul, haya, roble, álamo, cerezo y diversas variedades de pino) y esmerándose en trabajo y acabado.

On the left, above: coffee table in black poplar ref. 5044-105/03
130 x 80 x 51 cm, below: another in oak ref. 1240-003/03 130 x
75 x 54 cm. Below these lines: one with an original top
(different materials and sizes) and a pine one,
transparent glass and drawer ref. 7822-071/01.
A la izquierda, arriba: mesa de centro en chopo ref.5044-105/03
130x80x51cm, debajo: otra en roble ref.1240-003/03
130x75x54cm. Bajo estas líneas: una con original tablero
(diferentes materiales y tamaños) y una de pino,
cristal transparente y cajón ref.7822-071/01.

The antique-style furniture of GIARETTA is made in solid polar (parts of old wood) with the care and skill of the great artisans and cabinetmakers. Large chest, small table and bookcase from the HISTORIA collection.

Los muebles de estilo anticuario de GIARETTA están elaborados en madera maciza de álamo (partes de madera vieja) con el cuidado y buen hacer de los grandes artesanos y ebanistas. Arcón, mesita y librería de la col.HISTORIA.

On the previous page: large chest, model CORNELIA, small table, model ZACCARIA and bookcase, model CLELIA. Above: dining room from the ARTE POVERA collection, model ORCELLO, sideboard and bookcase ref. DS031, table ref. DS096 and chairs, model VENEZIA ref. D61I.

En la página anterior: arcón mod.CORNELIA, mesita mod.ZACCARIA y librería mod.CLELIA. Arriba: comedor de la col.ARTE POVERA, mod.TORCELLO, aparador y librería ref.DS031, mesa ref.DS096 y sillas mod.VENEZIA ref.D61I.

Metallic table with blue
glass top and wooden
chairs with original
structural details from
BORGIA RUANO.
Below: two types of
study desk and the
revolving chair from the
same manufacturer.
Mesa metálica con
tablero de cristal azul y
sillas de madera con
originales detalles
estructurales de
BORGIA RUANO.
Abajo: dos aspectos del
pupitre de trabajo y el
sillón giratorio de este
mismo fabricante.

On the left: bench in solid
mahogany,
model REGENCIA
ref. COC3-M and, below:
magnificent dining room table
with shaped legs finished
with multi-directional wheels,
model VICTORIANO,
oval and with two extensions
ref. ETCL-M.
Furniture produced
by KEEN REPLICAS.
A la izquierda: banco
en caoba maciza
mod.REGENCIA
ref.COC3-M y, abajo:
magnífica mesa de comedor
con patas torneadas
rematadas por ruedas
multidireccionales
mod.VICTORIANO, ovalada y
dos extensiones ref.ETCL-M.
Muebles producidos por
KEEN REPLICAS.

El Mueble
Natural **Furniture**

Office atmosphere protected by a folding screen. All the items of furniture belong to the J1 collection by JOHN KELLY, made in cherry and hazelnut wood. Natural oil and wax finish applied by hand.

protegido por un biombo. Todos los muebles pertenecen a la col.J1 de JOHN KELLY, realizada en madera de cerezo y avellano. Acabado natural a base aceites y cera aplicados manualmente.

This is furniture of a classical line but with a great concern for modern needs (taking advantage of space, care with the environment). On the right: high chest of drawers with door and three drawers, as well as: high bar chair.

Se trata de muebles de líneas clásicas pero con una gran preocupación por las necesidades actuales (aprovechamiento del espacio, atención al medio ambiente). A la derecha: cómoda alta con puerta y tres cajones, además: silla alta para barra de bar.

El Mueble
Natural **Furniture**

TALGÖ produces furniture with traditional shapes in pine from their Norwegian forests, the solid wood being whitened and oiled later, providing it with a smooth, long-lasting and easy-to-maintain shine.

TALGÖ produce muebles con formas tradicionales en madera de pino de sus bosques noruegos, la madera maciza tratada con un blanqueado y aceitada posteriormente, proporciona un brillo suave, duradero y de fácil mantenimiento.

Stephanie Bett 180

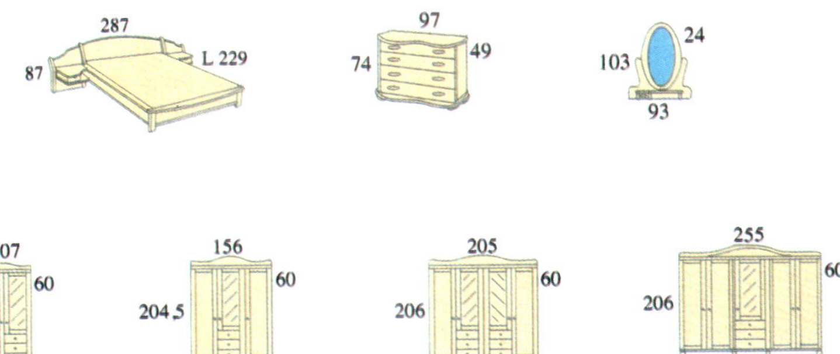

On these pages, the proposal is the STEPHANIE model bedroom: bed of 180 x 200 cm which includes the bedside tables, chest of drawers ref. 4401, mirror ref. 4483, and five-door wardrobe (the centre one with mirror). In the plans one can see smaller-sized wardrobes.

En estas páginas, propone el dormitorio mod.STEPHANIE: cama de 180x200cm que incluye las mesitas de noche, cómoda ref.4401, espejo ref.4483, y armario de cinco puertas (la central con espejo). En los esquemas se ve que hay armarios más pequeños.

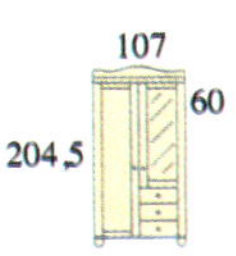

The furniture from KEEN REPLICAS is skilfully worked in solid mahogany and follows designs and guidelines of the period of its style. On the left, a colonial bed ref. BL-M, below a detail of the bedstead and bedside table, another in Imperial style ref. FLB2-M.

Muebles de KEEN REPLICAS trabajados artesanalmente en caoba maciza y siguiendo diseños y pautas de la época de su estilo. A la izquierda cama colonial ref.BL-M, abajo detalle del cabezal y la mesita, otra de estilo Imperio ref.FLB2-M.

On the following page: small dressing table with mirror and tiles ref. WVC-M, sideboard with larder in the upper part ref. KCC-M. Below: another two sideboards with the back decorated with tiles ref. WSCC-M and ref. WS-M.

En la página siguiente: mesita tocador con espejo y cerámica ref.WVC-M, aparador con alacena en la parte superior ref.KCC-M. Abajo: otros dos aparadores con trasera decorada con cerámica ref.WSCC-M y ref.WS-M.

Always basing its work on solid wood crafted in the traditional way, MYOC comes up with items of furniture which, designed for rustic atmospheres, are not put to shame in elegant surroundings. Alongside these lines: bookcase ref. 03. Siempre a base de madera maciza elaborada a la manera tradicional, MYOC propone unos muebles que, ideados para ambientes rústicos, no desmerecen en composiciones elegantes. Junto a estas líneas: librería ref.03.

Another model of vertical and stout bookcase ref. 08, bottle rack, open and closed ref. 06. On the following page: mixed wardrobe, open (closed, below, in the extreme right hand side of the page) ref. 106, another colonial wardrobe ref. 04 and large chests ref. 08 and 08T. Otro modelo de librería vertical y panzuda ref.08, arcón botellero, abierto y cerrado ref.06. En la página siguiente: armario mixto abierto (cerrado, abajo, en el extremo derecho de la página) ref.106, otro armario colonial ref.04 y arcones ref.08 y 08T.

In a solid, stately style
and with a delicately
smooth finish,
ANFE MUEBLES
presents its number 17
series with dining room
and lounge furniture.
Side adornments with
flutes, shaping,
openwork and
mouldings.
En un estilo sólido,
señorial y con acabado
delicadamente
patinado, ANFE
MUEBLES presenta su
conjunto nº17 con
mobiliario de comedor y
sala de estar.
Adornos laterales en
estrías, torneados,
calados y molduras.

*Armchair and chairs, model
ANTIQUE ref. 139 and 135,
glass cabinet with two doors
ref. 932, sideboard with three
doors and storage cupboard
(both 163 cm long) ref. 903
and 909, extendible table,
model ANTIQUE ref. 127 and
coffee table with glass
display top ref. 921.
Sillón y sillas mod.ANTIQUE
ref.139 y 135, vitrina con dos
puertas ref.932, aparador de
tres puertas con un altillo
(ambos con 163cm de largo)
ref.903 y 909, mesa
extensible mod.ANTIQUE
ref.127 y mesita de centro
con tablero expositor
en cristal ref.921.*

ARTEFERRO also produces solid wood furniture and classical lines. On the right, in column: writing desk with upper body and folding top ref. 1120, below: carved large chest and framed mirror with fanned crown ref. 2406 and 1459.
ARTEFERRO, también produce muebles en madera maciza y líneas clásicas. A la derecha, en columna: escritorio con cuerpo superior y tapa abatible ref.1120, abajo: arcón tallado y espejo enmarcado con remate en abanico ref.2406 y 1459.

Above these lines: secret writing desk with drawers ref. 1111, shelving, model HICKORY ref. 2738, work table with two drawers and shaped legs, model BLOOMINDALE ref. 1142 which is completed with a revolving lectern ref. 7162.
Sobre estas líneas: escritorio secreto con cajones ref.1111, estantería mod.HICKORY ref.2738, mesa de trabajo con dos cajones y patas torneadas mod.BLOOMINDALE ref.1142 que se completa con un atril giratorio ref.7162.

The furniture, presented by A.T. on these pages, is from the N.a.K. collection and blends all the advantages to be found from this manufacturer: select solid pine, painstakingly finished with waxing in a natural shade as well as large and practical drawers.

El mueble, presentado por A.T. en estas páginas, es de la col.N.a.K. y reúne todas las ventajas de este fabricante: selecta madera maciza de pino, esmerado acabado encerado en tono natural, además, amplios y prácticos cajones.

This sideboard and glass cabinet have original proportions of a slightly oriental look, backed up by the grille of wooden strips on the central body doors. The sides of the glass cabinet have bulging glass. Model MARLENE 268 x 70 x 211 cm.

Este mueble aparador y vitrina tiene unas originales proporciones de aspecto ligeramente oriental que se refuerza con el enrejado de maderitas en las puertas del cuerpo central. Laterales de la vitrina con cristal combado. Mod.MARLENE 268x70x211cm.

With a very traditional style, TORONDELL produces very original items of furniture thanks to the versatility in the distribution of drawers, shaped elements and mouldings. Below are featured the plans with other small table sizes.

Con un estilo muy tradicional, TORONDELL logra unos originales muebles gracias a la versatilidad en la distibución de cajones, elementos torneados y molduras. Abajo se aprecian los esquemas con otros tamaños de mesitas.

On this page: coffee and corner tables model 603 130 x 70 x 43 cm and model 605 60 x 60 x 43 cm, bar unit, model 116 66 x 46 x 90 cm. On the following page: a wardrobe with one door, two adjustable shelves and side drawers, model 120 92 x 44 x 155 cm.

En esta página: mesas de centro y rincón mod.603 130x70x43cm y mod.605 60x60x43cm, mueble bar mod.116 66x46x90cm. En la página siguiente: un armario ropero con una puerta, dos estantes móviles y cajones en el lateral mod.120 92x44x155cm.

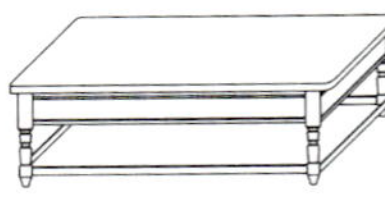
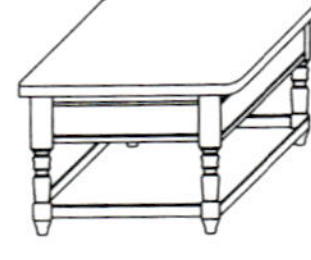
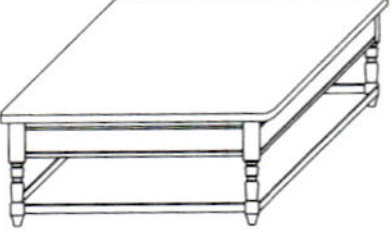
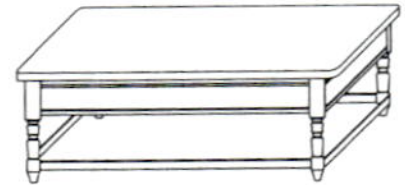
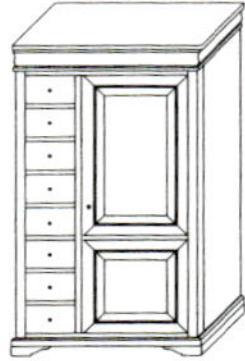

El Mueble
Natural **Furniture**

Natural furniture can be in traditional forms with thick, solid elements, or simple and elegant studied lines. On this double page ALBOR DESIGN proposes a modern atmosphere produced with its COMODÍN collection.

Un mueble natural puede ser de formas tradicionales con gruesos elementos macizos, o estudiadas líneas simples y elegantes. En esta doble página ALBOR DESIGN propone un moderno ambiente realizado con su col.COMODÍN.

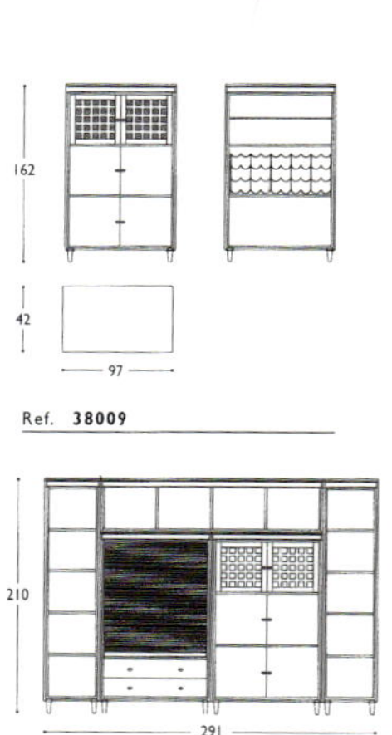

Ref. 38009

Ref. 38008 Hi-Fi

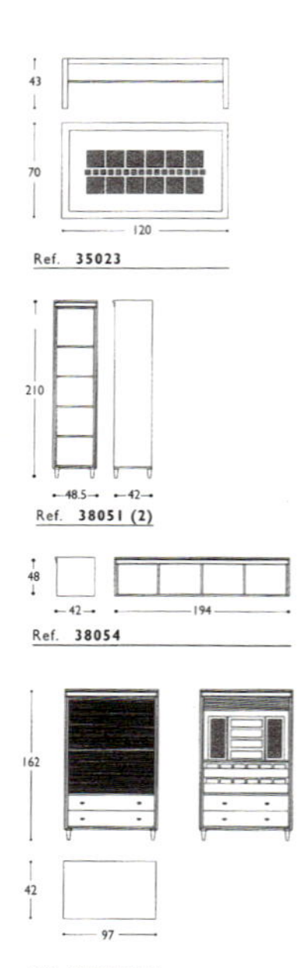

Ref. 35023

Ref. 38051 (2)

Ref. 38054

In the plans, one can see the composition of the modular furniture (with its references and measurements) which includes all that is necessary for a living room: shelves for books, special draws for music, glass cabinet, drawers, cupboards with doors or blind, etc.

En los esquemas, se aprecia la composición del mueble modular (con sus referencias y medidas) que abarca todo lo necesario para una sala de estar: estantes para libros, cajones especiales para música, vitrina, cajones, armarios con puertas o persiana, etc.

On this page, HABUFA proposes a dining room set from the LORE collection with finishes in solid aged fir and two models of larder with drawers and doors from the BRIGHTON collection ref. 1557 and ref. 1556 in waxed fir.
En esta página, HABUFA propone un conjunto de comedor de la col.LORE con acabados en abeto macizo envejecido y dos modelos de alacena con cajones y puertas de la col.BRIGHTON, ref.1557 y ref.1556 en abeto encerado.

On the left: small cupboards with different distribution of drawers and doors, OLD FINISH collection in solid pine with aged finish. Below: dining room from the BRIGHTON collection (solid waxed fir) whose table with two drawers ref. 1550 is delivered ready to assemble.

A la izquierda: armaritos con diferentes distribuciones de cajones y puertas, col.OLD FINISH en pino macizo y acabado envejecido. Abajo: comedor de la col.BRIGHTON (abeto macizo encerado) cuya mesa con dos cajones ref.1550 se entrega preparada para montar.

Furniture from R. JUSTE REQUENA in which the rustic elements of the setting do not conceal its great quality and deliberate traditional style: perfect for a country house. References and measurements in the plans.

Muebles de R. JUSTE REQUENA en que, los elementos rústicos del ambiente no ocultan su gran calidad y estilo voluntariamente tradicional: perfectos para una mansión en el campo. Referencias y medidas en los esquemas.

REF. # 7002
112 x 45 x 210 A. cms.
44¼ x 17¾ x 82¾ H. inches.

REF. # 328
185 x 41 x 204 A. cms.
73 x 16¼ x 80⅓ H. inches.

REF. # 5/H
Ø140 x 75 A. cms.
Ø55¼ x 29½ H. inches.

REF. # 5/H2
Ø110 x 75 A. cms.
Ø43¼ x 29½ H. inches.

REF. # 102/WS
49 x 45 x 107 A. cms.
19¼ x 17¾ x 42¼ H. inches.

Standing out, due to their elegance and charm, are the shaped legs and the wavy flap of the sideboard with shelves, the doors in trellis with wooden strips crossing diagonally and the solid structure with the shaped table (folding flaps) and chairs (raffia seat).

Destacan, por su elegancia y encanto, las patas torneadas y el faldón ondeado del aparador con estantes, las puertas en celosía con maderitas cruzadas en diagonal y la robusta estructura con torneados de la mesa (alas plegables) y las sillas (asiento de rafia).

On the right, from ARTEFERRO, small tables ref. 2286 and ref. 1102. BORGIA RUANO proposes furniture in which the naturalness of the lines is imposed but without foregoing a discreet decorative intention. Below, dining room with interesting outlined decoration in the sideboard.

A la derecha, de ARTEFERRO, mesitas ref.2286 y ref.1102. BORGIA RUANO propone muebles en los que se impone la naturalidad de líneas no exentas de una discreta voluntad decorativa, abajo: comedor con interesante decoración esgrafiada en el aparador.

*Below these lines: glass cabinet of wood and glass ref. 2512, from ARTEFERRO.
On the right: a cupboard model for TV, etc. with two carved doors. Below: tables with metal
structure and thick top (several sizes) from BORGIA RUANO.
Bajo estas líneas: vitrina de madera y cristal ref.2512, de ARTEFERRO. A la derecha: un modelo de
armario para TV, etc. con dos puertas y labrados. Abajo: mesas con estructura de metal y grueso
tablero (varias medidas) de BORGIA RUANO.*

EL MARANGON produces furniture in solid cherry or walnut, following the process and style of traditional cabinetmakers but adapting to the modern world. Hand-finished with shellac and beeswax.

EL MARANGON realiza muebles en madera maciza de cerezo o nogal. Siguen el proceso y estilo de ebanistería tradicionales pero adaptados al mundo actual. Acabados a mano con goma laca y cera de abejas.

Imperial lines, above: sideboard ref. 151C, rectangular extendible table ref. 450C, chair ref. 992CV and glass cabinet ref. 153C. Below: bedside table ref. 762C, bed ref. 656C, wardrobe ref. AC6IA, chest of drawers and mirror ref. 761C and 895C. Sideboards ref. 1138 and ref. 1120.

Líneas imperio, arriba: aparador ref.151C, mesa rectangular extensible ref.450C, silla ref.992CV y vitrina ref.153C, debajo: mesita de noche ref.762C, cama ref.656C, armario ref.AC6IA, cómoda y espejo ref.761C y 895C. Aparadores ref.1138 y ref.1120.

El Mueble
Natural **Furniture**

NAFARROA proposes solid items of furniture in traditional and rustic forms that do not ignore a skilful working and finish. A company specialising in customised jobs and full rehabilitation projects, including exact reproductions.
NAFARROA propone muebles macizos de formas tradicionales y rústicas que no impiden un cuidadoso trabajo y acabado. Empresa especializada en trabajos a medida y proyectos completos de rehabilitación, incluyendo reproducciones exactas.

This double page shows a wide selection of possibilities, always in fine wood that also includes ironwork or stone work: bedrooms, chests of drawers, sideboards, bookcases, glass cabinets, tables, rustic benches, console tables, etc.
En esta doble página, muestra una amplia selección de sus posibilidades, siempre en maderas nobles abarcando, también, trabajos en forja o piedra: dormitorios, cómodas, aparadores, librerías, vitrinas, mesas, bancos rústicos, consolas, etc.

DANVES proposes a selection of furniture that carries its name: the DANVES collection is produced using craftsmen's techniques and reproduces Spanish colonial furniture. Chest of drawers ref. 716, small tables ref. 711 and 708, sofa (den) model T/LEONARDO ref. 14.

DANVES propone una selección de muebles que lleva su nombre: col.DANVES, está elaborada artesanalmente y reproduce mobiliario colonial español. Cómoda ref.716, mesitas ref.711 y708, sofá (nido) mod.T/LEONARDO ref.14.

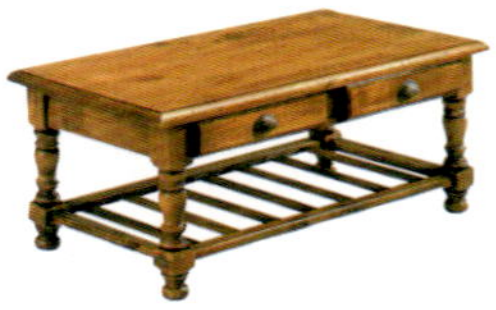

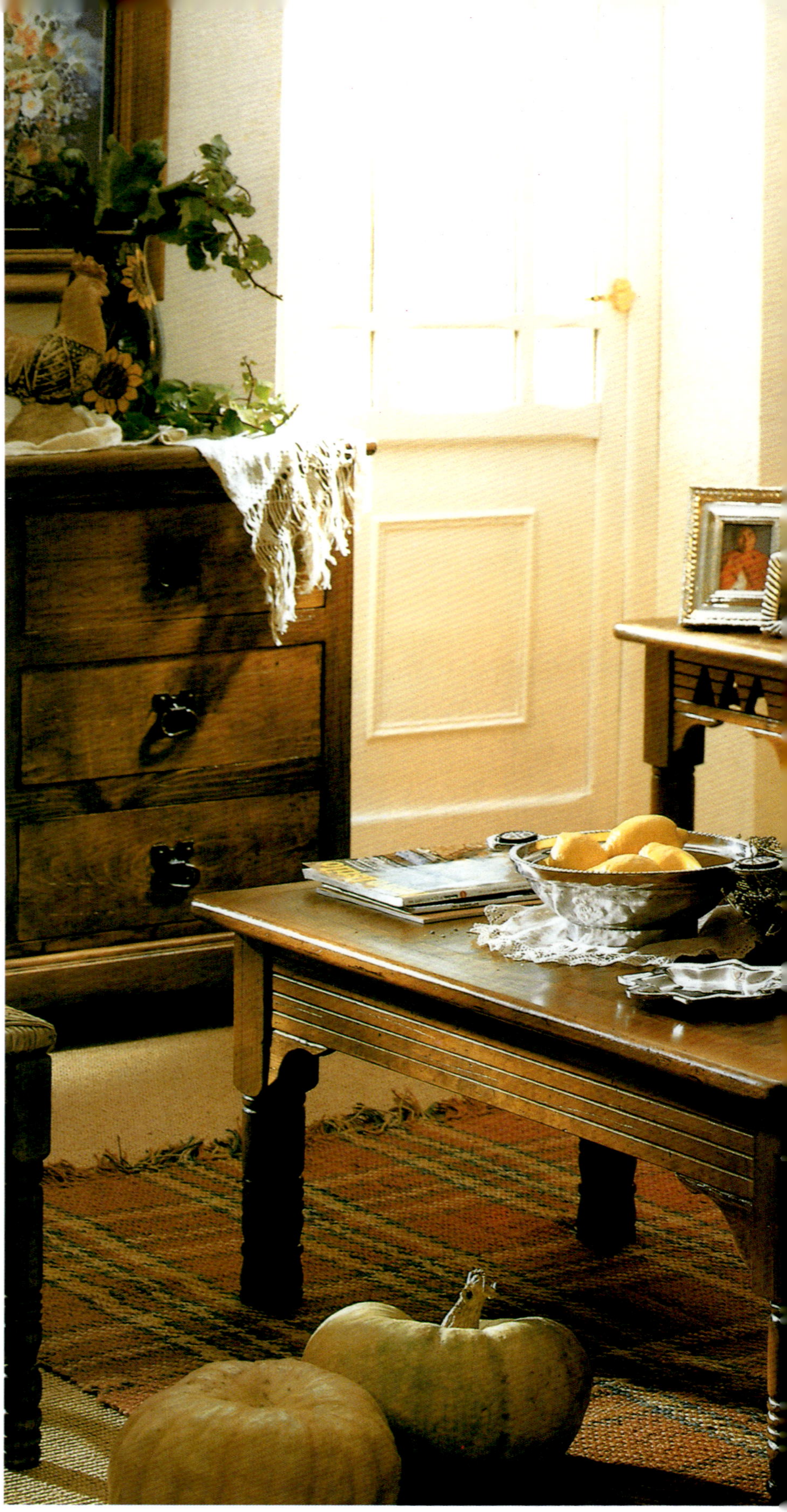

DANVES collection, left, from top to bottom: small tables ref. 420, 425 and 439. HACIENDA collection of two bookcases and bar unit (open and closed) ref. 540, 541 and 548. On the right, DANVES collection: small tables ref. 460, 479, and ref. 771 (rectangular and square).

Col.DANVES, izquierda, de arriba a abajo: mesitas ref.420, 425 y 439. Col. HACIENDA dos librerías y mueble bar (abierto y cerrado) ref.540, 541 y 548. A la derecha, col.DANVES: mesitas ref.460, 479, y ref.771 (rectangular y cuadrada).

TORONDELL proposes furniture made from the best pine and cherry and with a truly craftsman-like finish. This page shows a bedroom with three possible measurements for the bedstead (145/160/190x5x115cm).

TORONDELL propone muebles realizados con las mejores maderas de pino o cerezo y con un acabado verdaderamente artesanal. En esta página, presenta un dormitorio con tres posibles medidas para el cabezal (145/160/190x5x115cm)

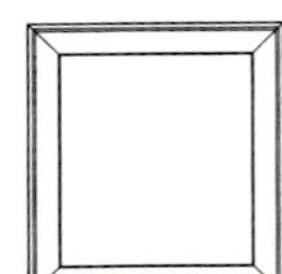

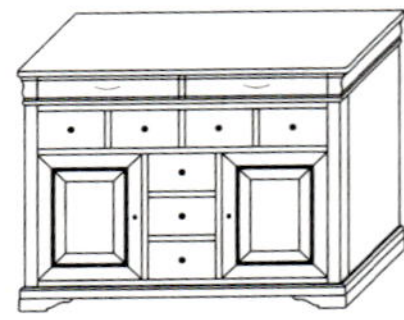

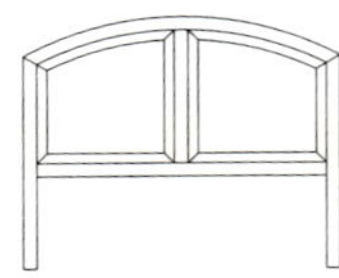

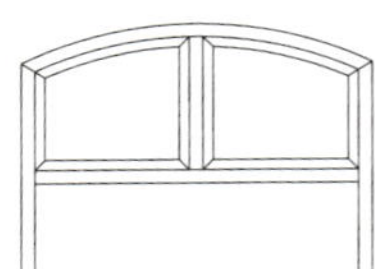

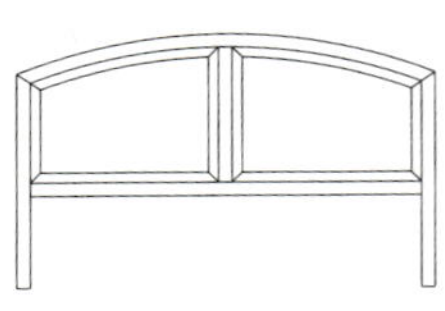

We come across: chest of drawers with framed mirror, model 104 118 x 46 x 90 cm and model 160-A 95 x 3 x 105 cm, bedstead model 150-C 160 x 5 x 115 cm and bedside tables with four drawers, model 131 50 x 39 x 63 cm.
The plans show the distribution of the drawers, etc.

Encontramos: cómoda con espejo enmarcado mod.104 118x46x90cm y mod.160-A 95x3x105cm, cabezal mod.150-C 160x5x115cm y mesitas de noche con cuatro cajones mod.131 50x39x63cm. En los esquemas se ve la distribución de los cajones, etc.